I0824572

Alight

How Urban Parrots Found New Places to Belong

Written by
Jennifer Torres

Illustrated by
Molly Magnell

atheneum
Atheneum Books for Young Readers
New York Amsterdam/Antwerp London Toronto Sydney/Melbourne New Delhi

In golden California,
as the sky turns pink,
parrots awake.

By the thousands.

They squawk.
They screech.
They swoop—

a flash of green, a splash of red—
from palm trees to power lines.

Cackling, they swipe fruits from tidy gardens. Cawing and clamoring, they flap above pools and parking lots.

They alight on branches where parrots do not belong. And yet, they are here. At home among traffic lights and train stations.

And perhaps, as you peer up into raucous treetops—as tropical flocks take wing in unlikely skies—you wonder, *Where did all these parrots come from?*

The parrots came from someplace else. That is the simple truth of it. That is a *fact.*

But a fact is not a full story. And the parrots' story, like so many true stories, is not so simple.

Once, the parrots came from Mexico.
They belonged among rugged slopes and
wooded canyons along the coast.

They foraged fruits and nuts, flowers and seeds.
They raised their chicks in leafy treetops, twittering among the huisache and strangler fig.

But over time,
their forest homes
were destroyed.

Their nestlings were stolen.
They were bundled away
to distant cities
and sold as pets.

By the thousands.

By the tens of thousands.

Some of those pets were turned loose
from their cages.

(Parrots are noisy.
They will not be hushed.)

Others escaped,
like wild things do.

And then those daring runaways took flight over golden California, where so many newcomers had ventured before them.

After that, they might have wandered, lonely and lost.
Far from home, they might have dwindled.
Even disappeared.
Their story might have ended.

But it did not.

Because, instead of rugged slopes,
the parrots learned to soar above shaded sidewalks.

Instead of wooded canyons,
they made their homes among
bus stops and apartment buildings.

Searching for familiar fruits and flowers, they found gardens to rival forest feasts: oranges and avocados, persimmons and figs. (These fruit trees are from someplace else too.)

And, most important of all, they chirped and chattered and found one another.

In Mexico, where the parrots first came from,
there are few of them left. And every day fewer.

But, elsewhere, they have found new places to belong.
And when the sky fades again to purple,
they return to their roosts.

They squawk.
They screech.
They survive.

They thrive.

"**We are here!**" they seem to cry out, bold and brave and boisterous. (Parrots, as you know, will not be hushed.)

"**We are home!**"

Author's Note

Red-crowned parrots make noisy neighbors. Just before sunrise and again at dusk, they cackle on the utility lines outside my kitchen window in Southern California—a place where parrots did not always belong. Their squawking and screeching tell a story of survival, adaptation, and the complicated relationship between humans and the environment.

Red-crowned parrots—or in Spanish, loros tamaulipecos—are native to the foothills of Northeast Mexico. (Their Spanish name refers to the Mexican state of Tamaulipas, part of the bird's natural **habitat**.) So, what are wild flocks doing over suburban Los Angeles?

According to urban legend, they are the descendants of parrots that once escaped a pet store fire or that broke free from their cages at the Busch Gardens theme park that used to be located in Los Angeles. There is probably an element of truth to both stories. But here's what ecologists—scientists who study **ecology**—believe *really* happened:

From the 1960s through the 1980s, tens of thousands of red-crowned parrots were taken from their roosts in Mexico and brought to the United States—both legally and illegally—to be sold as pets. Some of those birds escaped. Others were released, perhaps by flustered owners who weren't prepared for the racket they make.

Often, **introduced** species do not survive outside their native **range**. But red-crowned parrots are resilient. They found new sources of food in the fruit and nut trees (also introduced) that Southern Californians love to plant in their backyard gardens. The social birds also found *one another* and formed thriving new colonies.

By 2001, the red-crowned parrot had become such a part of Southern California's **ecosystem** that the California Bird Records Committee added it to the official list of state birds. Today, scientists estimate that there are as many as three thousand red-crowned parrots flying wild in the skies above Los Angeles—a population that may now outnumber those found in Mexico, where poaching and the loss of forest habitat to logging, agriculture, and climate change threaten the birds' survival. The International Union for Conservation of Nature classifies the red-crowned parrot as **endangered**.

If their numbers continue to dwindle in Mexico, the species' future may depend on satellite flocks living in California and elsewhere. (Wild red-crowned parrots have also made homes in Texas, Florida, Hawai'i, and Puerto Rico.) This has led some scientists to think about **conservation** in new ways.

Introducing plants and animals into places where they are not naturally found raises important concerns. Such species may turn out to be **invasive**, damaging the ecosystem by crowding out native plants and animals, or by competing for scarce food supplies. But this doesn't seem to be the case with the red-crowned parrot in California. Now, ecologists wonder if careful study and monitoring of the red-crowned parrot might make it possible to help other endangered species find new places to belong.

Urban Parrots Everywhere

California isn't the only state that is home to flocks of wild parrots. In 1992, the US government passed the Wild Bird Conservation Act to help protect wild bird populations around the globe. But until then, the United States was the world's largest importer of wild parrots. According to the World Wildlife Fund, approximately 150,000 parrots were imported to the United States in 1990 alone. Now, scientists have identified fifty-six different parrot species living in forty-three states and Puerto Rico. The brightly colored birds are increasingly common in cities from New York to Miami and Chicago to Los Angeles. Like the red-crowned parrot, most were introduced through the pet trade. Have you spotted any wild parrots where you live?

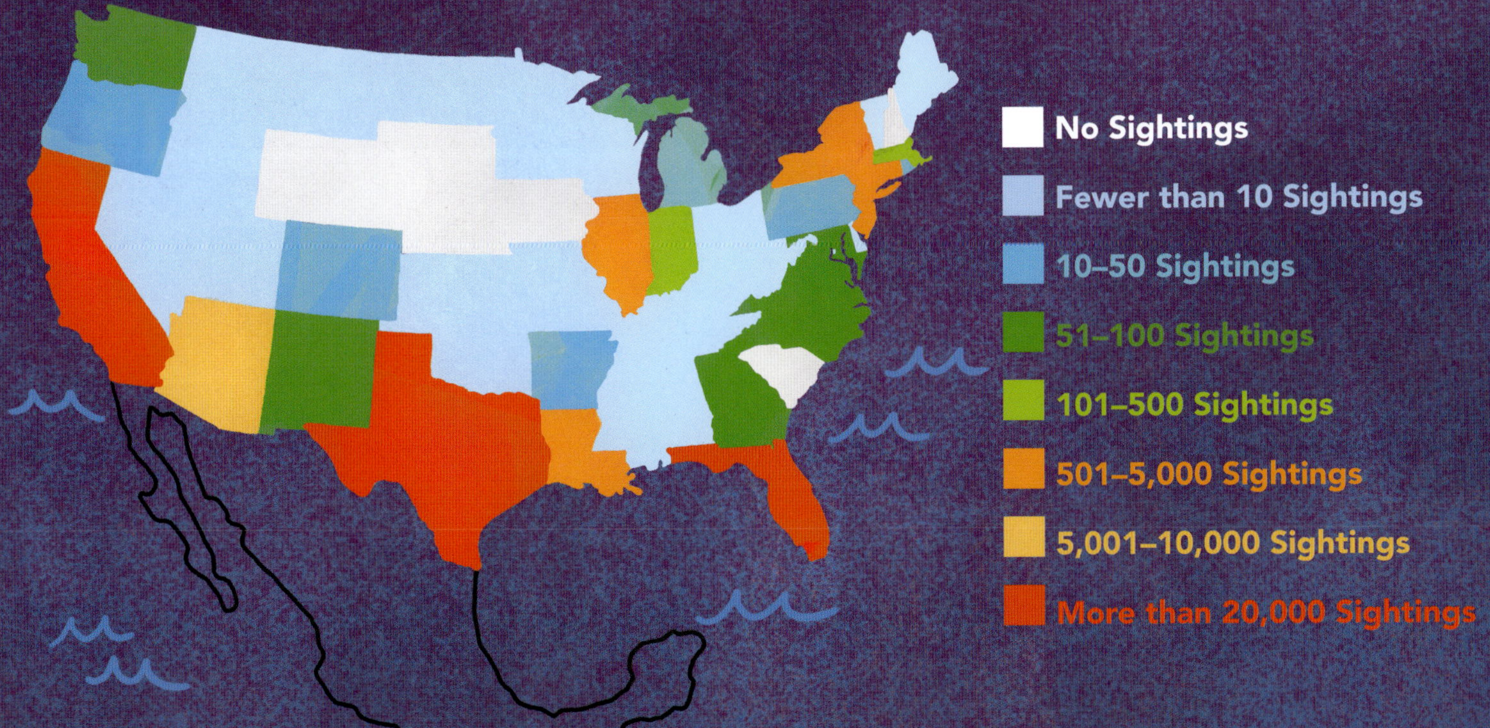

Wild parrots have been spotted in nearly every state, but they are most commonly seen in California, Florida, and Texas. (Hawai'i and Puerto Rico, which also have large populations of wild parrots, are not shown on this map.) Researchers at Cornell University, University of Michigan, and University of Chicago analyzed bird sightings reported by citizen scientists from 2002 to 2016 to understand where and how many wild parrots are living in the United States. A parrot sighting does not always mean that parrots will survive and establish large populations in that state, the way the red-crowned parrot has in Los Angeles. Many factors—especially weather and food availability—determine whether the birds will be successful in new environments.

Other Urban Parrots

Monk Parakeet: The monk parakeet—native to Bolivia, Brazil, and Argentina—is found in twenty-five states and the most commonly seen parrot in the United States. Unlike most parrots, which nest in existing cavities (small holes or crevices), monk parakeets build their own nests. Scientists believe these cozy nests are one reason the birds have been able to survive even in northern states—including New Jersey and New York—where they rely on backyard bird feeders for food throughout the cold winter months. However, monk parakeets often build their nests near utility poles and power lines, which can cause electrical outages and even fires.

Rose-Ringed Parakeet: Native to South Asia and Africa, rose-ringed parakeets are now found in at least forty-seven countries around the world, making them one of the most widespread parrot species. In Hawai'i, where rose-ringed parakeets were brought as pets in the 1960s, wild flocks have caused damage to fruit and grain crops.

Yellow-Headed Parrot: These clever birds are popular pets because they can be trained to talk. This may be one reason they are endangered in their native habitats of Mexico, Guatemala, Honduras, and Belize. Like the red-crowned parrot, yellow-headed parrots have established flocks in California, Florida, and Texas.

Rosy-Faced Lovebird: Thousands of these small, chirpy parrots, native to southwestern Africa, have made new homes in Arizona, where they nest in saguaro cactuses and perch on air-conditioning vents to keep cool.

Nanday Parakeet: Like other wild parrots in the United States, Nanday parakeets—native to Brazil, Paraguay, and Argentina—often make their homes in urban and suburban areas, where they can find nests and food in human-built environments. However, some Nanday parakeets have found ways to thrive outside of cities in California's Santa Monica Mountains.

Those Places Look Familiar

Explore some of the real places that inspired Molly's illustrations!

Hollywood Sign: Wild parrots—and their chatter—have become almost as familiar to the Southern California landscape as the iconic Hollywood sign, built in 1923. And, like the sign, flocks of red-crowned parrots make Jennifer think of Los Angeles as a city full of strivers, chasing dreams and new beginnings.

Union Station: Palm trees, such as the ones that stand tall outside Union Station in downtown Los Angeles, are favorites among red-crowned parrots, which nest in cavities in the trees. Just like the parrots, these trees were brought to the region from elsewhere.

Tamaulipas, Mexico: In northeastern Mexico, the red-crowned parrot's native range, researchers and volunteers are working to protect local bird populations and their nesting sites, which have been weakened by habitat loss and the illegal pet trade.

Santa Monica Pier: Thirteen species of wild parrots soar in the California skies. Birders have even spotted some in the beach neighborhoods surrounding the Santa Monica Pier.

The Huntington: Red-crowned parrots are frequent visitors at botanical gardens, including those at the Huntington in San Marino, California, where fruit trees and fountains offer them food and shelter. The raucous birds are especially fond of the Huntington's palm- and subtropical-themed gardens.

Terms to Know

Conservation: The protection and preservation of plants, animals, and other natural resources.

Ecology: The study of the relationship between living things and their surrounding environment.

Ecosystem: The interconnected community of living things—including plants and animals, as well as nonliving materials like soil and rocks—in a particular area.

Endangered: A plant or animal species is endangered when it is at risk of dying off completely and becoming extinct.

Habitat: The place where a plant or animal lives. For an animal to survive in a habitat, it must be able to find food, water, and shelter there.

Introduced: When a plant or animal species is brought into an area where it isn't naturally found. Species can be introduced on purpose or by accident.

Invasive: A plant or animal species that causes harm or damage when it is brought into an area where it isn't naturally found. Not all introduced species are invasive.

Range: The geographical area where a species is naturally found.

Additional Resources

The California Parrot Project *https://CaliforniaParrotProject.org*
This scientific organization is dedicated to studying and monitoring the thirteen parrot species currently found in California, and to educating the public about parrots in urban ecosystems.

American Bird Conservancy *https://abcBirds.org*
This organization works to protect wild birds and their habitats throughout the Americas. Working together with the Mexican organization Pronatura Noreste (*https://PronaturaNoreste.org*), the American Bird Conservancy hopes to help local communities preserve the red-crowned parrot's natural habitat in Tamaulipas, Mexico.

eBird *https://eBird.org*
This citizen-science project, managed by the Cornell Lab of Ornithology, invites visitors to explore photos and audio recordings of red-crowned parrots and other wild birds. Discover where wild parrots have been sighted, and take action by submitting your own observations!

Selected Sources

Calzada Preston, Carlos E., and Stephen Pruett-Jones. "The Number and Distribution of Introduced and Naturalized Parrots." *Diversity* 13, no. 9 (2021).

Conroy, Gemma. "Exotic Parrot Colonies Are Flourishing Across the Country." National Audubon Society, June 5, 2019.

Cornell Lab of Ornithology. "Saving Our Shared Birds: Partners in Flight Tri-National Vision for Landbird Conservation." 2010.

Curtiss, Aaron. "Question of How Wild Parrots Flew the Coop Is Up in the Air." *Los Angeles Times*, 1991.

Garrett, Kimball L. "Introducing Change: A Current Look at Naturalized Bird Species in Western North America." *Western Field Ornithologists—Studies of Western Birds*, 2018.

Gómez de Silva, Héctor A., Adán Oliveras de Ita, Jorge G. Álvarez-Romero, Clementina Equihua, and Rodrigo A. Medellín. Amazona Viridigenalis. "Vertebrados superiores exóticos en México: diversidad, distribución y efectos potenciales." Instituto de Ecología, Universidad Nacional Autónoma de México, 2005.

Heise, Ursula K. *Urban Ark Los Angeles*. KCET, 2018.

Natural History Museum of Los Angeles County. *Curiosity Show* 9: "Flocks Over Los Angeles." 2016.

Shaffer, H. Bradley. "Urban Biodiversity Arks." *Nature Sustainability* 1 (2018): 725–727.

Uehling, Jennifer J., Jason Tallant, and Stephen Pruett-Jones. "Status of Naturalized Parrots in the United States." *Journal of Ornithology* 160, no. 3 (2019): 907–921.

For David
—J. T.

For Bonnie, Ana,
Lulu, and Suzu
—M. M.

ATHENEUM BOOKS FOR YOUNG READERS • An imprint of Simon & Schuster Children's Publishing Division • 1230 Avenue of the Americas, New York, New York 10020 • For more than 100 years, Simon & Schuster has championed authors and the stories they create. By respecting the copyright of an author's intellectual property, you enable Simon & Schuster and the author to continue publishing exceptional books for years to come. We thank you for supporting the author's copyright by purchasing an authorized edition of this book. • No amount of this book may be reproduced or stored in any format, nor may it be uploaded to any website, database, language-learning model, or other repository, retrieval, or artificial intelligence system without express permission. All rights reserved. Inquiries may be directed to Simon & Schuster, 1230 Avenue of the Americas, New York, NY 10020 or permissions@simonandschuster.com. • Text © 2026 by Jennifer Torres • Illustration © 2026 by Molly Magnell • Book design by Lauren Rille • All rights reserved, including the right of reproduction in whole or in part in any form. • ATHENEUM BOOKS FOR YOUNG READERS is a registered trademark of Simon & Schuster, LLC. Atheneum logo is a trademark of Simon & Schuster, LLC. • For information about special discounts for bulk purchases, please contact Simon & Schuster Special Sales at 1-866-506-1949 or business@simonandschuster.com. • Simon & Schuster strongly believes in freedom of expression and stands against censorship in all its forms. For more information, visit BooksBelong.com. • The Simon & Schuster Speakers Bureau can bring authors to your live event. For more information or to book an event, contact the Simon & Schuster Speakers Bureau at 1-866-248-3049 or visit our website at www.simonspeakers.com. • The text for this book was set in Avenir. • The illustrations for this book were rendered in pencil with digital painting. • Manufactured in China • 0226 SCP • First Edition
10 9 8 7 6 5 4 3 2 1
CIP data for this book is available from the Library of Congress. • ISBN 978-1-6659-3877-8 • ISBN 978-1-6659-3878-5 (ebook)